Collected Papers on General Telecommunications Theory

A Systems Approach

Submitted to Northwestern International University as a Requirement for the Doctor of Philosophy Degree in Telecommunications.

by

Bradley S. Tice

authorHOUSE

1663 LIBERTY DRIVE, SUITE 200
BLOOMINGTON, INDIANA 47403
(800) 839-8640
www.authorhouse.com

First published by AuthorHouse 08/04/04

ISBN: 1-4184-1769-6 (e)
ISBN: 1-4184-1770-X (sc)

Printed in the United States of America
Bloomington, Indiana

This book is printed on acid-free paper.

Northwestern International University

Ltd.

NIU

Cyprus

Be it known that

Bradley S. Tice

having completed the course of study as prescribed by the Faculty and Board of Trustees, and having complied with all other requirements of the University is awarded the Degree of

Doctor of Philosophy in Telecommunications

In Testimony Whereof the Board of Trustees upon recommendation of the Faculty has granted this diploma bearing the seal of the University Dated at the University in Limasool, Cyprus.

July 15, 2003

Chairperson of the Board of Trustees

President

Dean of Students

In Memoriam
Hedwig Eva Maria Kiesler
(A.K.A. Hedy Lamarr)*
(b. November 9, 1913 – d. January 19. 2000)

*Known for her beauty, actress Hedy Lamarr was a co-inventor, with her then husband, the musician George Antheil, of 'Frequency-Hopping Radio' in their patent on secret communication systems in 1941. Granted an American patent in 1942, it is currently being used in what is known as spread-spectrum technology that is used in a wide array of electronic products.

H.J. Braun (1997) "Advanced weaponry of the stars". Invention & Technology, Volume 12, Number 4, Spring 1997, pp. 10-16.

D.R. Hughes and D. Hendricks (1998) "The improbably inventors of frequency-hopping radio". Scientific American, Volume 278, Number 4, April 1998, pp. 95.

A.L. Unterburger (1998) International Dictionary of Film and Filmmakers-3: actors and Actresses. New York: St. James Press, pp. 669-671.

E. Weise (1997) "Sultry actress finally gets due for invention". San Jose Mercury News, March 10, 1997, pp. 1A and 8A.

C. Rickey (2000) "Hedy Lamarr, screen beauty". <u>San Jose Mercury News</u>, January 20, 2000, pp. 5B Obituaries.

"Hedy Lamarr". <u>The Economist</u>, January 29th-February 4, 2000, pp. 101, Obituary.

Abstract

Abstract

A compilation of telecommunications papers delivered in the years 1998 and 2000. They deal with general telecommunication theory with a focus on a systems approach.

.

Preface

Preface

My collected papers on general telecommunication theory have been compiled for my dissertation for my Doctor of Philosophy degree in Telecommunications. These papers were delivered in the years 1998 and 2000 and represent a systems approach to telecommunication theory.

Acknowledgement

Acknowledgement

I would like to thank The Tech Museum of Innovation Awards Committee for accepting my submission for the Economic Development category for my work on AIRNETS. Although I was not selected as a finalist, I was honored to be considered for this fine award.*

*The Tech Museum of Innovation
 201 South Market Street
 San Jose, California 95113-2008

Introduction

Introduction

This dissertation is a compilation of my collected papers on general telecommunication theory with a focus on a systems approach. The first paper was delivered in 1998 at a conference sponsored by the American Institute of Aeronautics and Astronautics (AIAA). The remaining two papers were given at the International Society for Optical Engineers (SPIE) conference in the year 2000.

While this is a collection of papers, each paper stands on its own and represents a separate chapter, with some blending of information as a whole due to the common nature of the paper's topics. They represent my technical knowledge and engineering vision for the field of telecommunications.

Chapter 1

Telecommunications

"NETSAT: The Use of Telecommunication Satellites for Networking Computer Systems"

A paper presented at the 1998 American Institute of Aeronautics and Astronautics (AIAA/AAS) Astrodynamics Specialist Conference, August 10-12, 1998, Boston Park Plaza, Boston, Massachusetts.

#98-4378

NETSAT:
THE USE OF TELECOMMUNICATION SATELLITES FOR NETWORKING COMPUTER SYSTEMS

Bradley S. Tice, Director
Advanced Human Design
Cupertino, California

Bradley S. Tice

Abstract

The use of current telecommunication technologies in designing and implementing a satellite network for various computer systems. Technical details will be taken from current thinking in research and be applied to a future plan of a global network of computer systems. From an overview of current technology a model of a global networking system will be designed that has national as well as international importance.

Introduction

Last year, 1997, was both the 50th Anniversary of ACM, Association of Computing Machinery, and the transistor, it is fitting that the two should meet, abet in a more developed form, in the concept and application of telecommunication satellites for a global computer network. Although the technology described for this process is current, the ideas behind such a communication relay system dates to the last year of World War Two and Arthur C. Clarke's seminal article "Extra-Terrestrial Relays" that proposed synchronous satellites to relay signals to and from earth (McAleer, 1987: 266 and Bruno, 1997: 226).

Several factors play into developments of satellites beyond the 'visionary' stage. It is clear that the great technology derived from the research carried out during World War Two, radar, atomic research; The Manhattan Project, that lead to the development of the transistor in 1947 by Shockley, Bardeen and Brattain. Silicon Valley is a direct by product of the 'transistor revolution' and the chronological list of its prodigy; the diffused based high frequency broadband amplifier of 1957, the microchip of 1967, the integrated circuit of 1971 and the digital signal processor chip of 1997 (Gillmor, 1997: E1 & E7).

It is interesting to see that the development of "The Web" has had some positive effect on increasing the sharing of scientific information with the general public (Dagani, 1997:70). As the writer K.K. Campbell states "Science is nothing but content, and content is the lifeblood of the internet. Since content is an increasingly endangered on-line species, science sites are becoming more prominent almost by default" (Dagani, 1997:70). Although an oversimplification, as science is not content as much as it is 'results in context', the need for an educated use of the internet must be a priority or it could stagnate into an over commercialized reflection of what is 'low' in popular culture.

On the other end of the spectrum is the 'over estimation' of such a system. Such books as <u>Darwin Among the Machines: The Evolution of Global Intelligence</u> by George B. Dyson (New York: Addison-Wesley, 1997), reviewed by Henry Rzepa in <u>Chemical & Engineering News</u>, gives the faint impression that more data makes for more information, i.e.. that pieces of fact, data, when collected and interpreted through multiple filters, the networking system, will produce the necessary energy to produce information, something with a perceptual content and context to some specific problem or problems (Rzepa, 1997: 71-74). Unfortunately, not only is this system compared to the human brain, but it is expected to exceed the human brain by a multiple of factors.

The internet is just a 'jumped up telephone system' wired to the neighbor's personal computer. It is an important development, but more along the lines of a systems operation, rather than a complex conceptual one. Communication is not necessarily always intelligent communication, as can be seen by the viewing of the average American television broadcast.

<u>Satellites</u>

In looking back at what was dreamed would be the future for satellites, and telecommunications as a whole, such items as: (Seifert, 1959; 2-27)

1.) Military needs such as missile delivery and reconnaissance.
2.) Communication Systems.
3.) Weather prediction.

Early predictions to the future of satellite use was mixed with a bevy of complex and unknown issues that presented themselves to the technical designers of such satellites. Pierce makes a point that although two systems have been proposed, Pierce's in 1955, and Arthur Clarke's 'synchronous satellites' in 1945, that relay satellites will only become useful if they perform for a greater length of time, i.e. meaning years of operation (Grey and Grey, 1962:27). Pierce is also clear in how he defines such future uses of such technology:

> I predict a brilliant future for satellite communications as the first point of union between space technology and the other fields of technology and science. Space technology has not yet made a close and practical alliance

with the technology that built today's world. Perhaps the communication field can bridge this gap (Grey and Grey, 1962:30).

In looking at reports from the early 1980's, the future of global telecommunications was shaping up to provide service for all regions of the world.

From the 1981 International Telecommunication Union Conference held in Geneva, Switzerland (International Telecommunications Union, 1981).

A major world investment in communication satellites, and much planning for future communication development, concerns the fixed satellite service (International Telecommunication Union, 1981:29).

Such conferences were important in regulating the future types of services to be provided for each country. Constraints of an international nature will be defining the direction such communication services will be taking in the future.

In reading some general secondary works on satellites it becomes clear that even as late as twenty years ago the idea of using satellites to relay all types of data around the world was being considered. Martin points out in his <u>The Wired Society</u> (1978) that the total investment of terrestrial telecommunications facilities is 'huge'. He states "AT&T alone is spending $9 to $10 billion per year on capital improvement of the Bell system" (Martin, 1978:149). He continues "The annual revenue from telecommunications in the United States is over $40 billion and is growing by about $5 billion per year. Much of the capital expenditure in the telecommunications industry is going into trunks and trunk switching that satellites and demand-assignment equipment could replace" (Martin, 1978:149).

Martin makes it clear that such uses can have tremendous cost savings. "If earth stations were associated with the toll offices in the 500 most populous cities, such a satellite network would cost less that $2 billion. If it had a lifetime of ten years, the cost would be $200 million per year. (Most telecommunications equipment has a forty-year lifetime.)" (Martin, 1978:150).

It is clear from the general investment in existing ground facilities, usually public works or private utilities companies that are paid for and maintained by tax dollars or direct charges, of the developing internet, such long term considerations were either ignored or forgotten when designing the current land based systems.

Telecommunications

With the rapid rise of a global economy, the explosive growth of the telecommunications industry will have a global impact in all aspects of our lives. In looking back a decade ago, the expected growth of international telephone traffic would double from the period 1983 to 1988 and double again from 1988 to 1993 (Shipman, 1987: 83).

With this increased demands comes an increase in efficiency of research, design, manufacture, and operation of such satellite systems. Because all communication satellites must be in the same geo-synchronous orbit that circle the earth's equator once every twenty-four hours, there is only a limited number of satellites that can be placed in that orbit, and are regulated by the

World Administrative Radio Conference, WARC (Shipman, 1987:83).

The way technology is transferred from government research to private industry is through what is called Space Technology Integration or STI. As Smith explains from his <u>Teleservices Via Satellite</u> (1978) "Space Technology Integration is a specific application of the basic TI framework which is relevant to the communication satellite experimental process. There are two major cycles in the process-the innovational and institutional cycles. The two cycles meet at the point where a specific application has reached the stage when it is ready for operational implementation. At this point the application goes on to the institutional cycle, and the innovational cycle begins again", (Smith, 1978:2).

Such a cycle can be seen with such organizations as NASA. In 1971 during a reorganization of its Office of Space Science and Applications, a new office was established, the Office of Applications (Smith, 1978:18). The central idea behind this new office was explained by then NASA's Administrator James C. Fletcher "The application of space technology to solving problems here on Earth is perhaps NASA's most important new thrust. I believe it appropriate to centralize into a single office in NASA Headquarters all of the resources which we can muster to support space applications to Earth" (Smith, 1978:18).

Current Developments

An overview of current technology gives a glimpse of the potential development of a satellite based distributional system for a global computer network. From the development of wireless modems that weigh but ounces to the rapid growth of satellites based mobile phones, the trend is towards a wireless world (Bluck, 1998:4).

One of the most important factors in driving this new market is the desire for more 'speed' in relation to a systems carrying capacity. Such a carrying capacity is a function of the power of the signal, the bandwidth of the signal, and the amount of noise that is present (Hogan, 1997:37). This is termed 'Shannon's Theorm' and determines how much data a medium can carry.

Megabit per second transmissions are only currently possible in three formats: cable, xDSL, and satellite (wireless). A trouble area for the development of a satellite Internet system is the problem of distance in regards to the Internet TCP/IP protocol that holds each sent packet until an acknowledgment is received of a successful transmission (Hogan, 1997:37). For a satellite in a geosynchronous orbit, 22,000 miles overhead, the delay for the round trip is half a second.

One solution is to drop the requirement for the 'acknowledgment' thus removing a extraneous part of the protocol. Another is by adding more memory to the buffer (Hogan, 1997:51). Deregulation of the radio frequencies have also contributed to developments such as local multipoint distribution service or LMDS that have a lower maintenance costs than copper or cable-based networks (Garber, 1998:14).

This cost savings also applies to satellite applications to the internet. There are 'substantial' savings in utilizing the Internet satellite technology as opposed to relatively expensive high-bandwidth terrestrial systems (Kilaski, 1997:52). This will be important in the near future as the internet is expected to grow from 50 million subscribers today to 150 million by the year 2000 with a growth of $30 billion in the ensuing years (Evans, 1998: 76). With the expected investment of around $400 billion for future Q/V-band systems, wave lengths of six to eight millimeters, that are technically more difficult than the currently proposed Ka-band that encompasses wavelengths between 1 and 1.5 centimeters that have serious attenuational problems with precipitation (Evans, 1998:76).

When discussing satellite orbits, there are three categories of orbits based on the distance from the earth. Closes to proximity to the earth is the LEO satellites, for Low Earth Orbit, that circle hundreds of kilometers above the surface of the earth. Next are the MEO satellites that represent a Medium Earth Orbit and at 36,000 kilometers out from earth are the GEO satellites (Pelton, 1998:81).

As the market gears towards a consumer based economy in the future, space-based systems will increasingly shift from commercial to consumer and compete with a bevy of large carrier based services. What does this mean to the Internet and the satellites used to relay the uplinks and downlinks of data for the new millenniums information age? The first answer is basic physics. The development of a finite resource will drive technology forward and spread that technology on a global scale. The second will be a world market that will move from a commercial oriented to a independent consumer based economy that will focus on the individual in the context of a global networking systems.

Some implications to current computing would be the following:

1.   A new view of the architecture of computing systems.
2.   Access to large parallel computers or 'super computers' that would otherwise be in the exclusive domain of academic or private research institutes.
3.   A true 'globalization' of computer systems that would span all areas of the earths surface.

Time will only tell how these systems develop in the future.

Summary

It is clear that the advantages of telecommunication satellites in relaying the growing volume of data around the world will far outstrip its terrestrial twin in both efficiency and volume on a global scale in the coming years ahead. What this means is that utilization of networking

(NET) satellites (SAT) that will form a global platform (NETSAT) to provide a global network for computer systems around the world.

References

Bluck, J. (1998) "Ames offers wireless modems slightly bigger than pagers" The Ames Astrogram, March 20, 1998, page 4 & 8.

Bruno, L.C. (1997) Science & Technology Firsts. New York: Gale.

Dagani, R. (1997) "Keeping up with science news on the web" in Chemical & Engineering News. November 24, 1997 Volume 75 Number 47, page 70.

Evans, J.V. (1998) "New satellites for personal communications". Scientific American. April 1998. Volume 278, Number 4, page 70-77.

Garber, L. (1998) "Auction sets stage for new broadband wireless service". Computer May 1998 Volume 31, Number 5.

Grey, J. and Grey, V. (1962) Space Flight Report. New York: Basic Books, Inc.

Gilmor, D. (1997) "Seeds of a revolution" San Jose Mercury Newspaper. Section E, page 1 and 7, December 21, 1997

Guthery, S. (1997) "Wireless relay networks" in IEEE Network November/December 1997 Volume 11, Number 6, pages 46-51.

Hogan, H. (1997) "Paving the information driveway" in High Technology Careers Magazine. October/November 1997 Volume 14, Number 7, pages 19, 37, 51.

International Telecommunication Union (1981) Twentieth Report by the International telecommunication Union on telecommunication and peaceful uses of outer space. Geneva: International Telecommunication Union.

Kilarski, D. (1997) "Satellite networks: Data takes to the skies" in Network December 1997 Volume 12 Number 13, pages 52-58.

Martin, J. (1978) The Wired Society. Englewood Cliffs: Prentice Hall, Inc.

McAleer, N. (1987) The Omni Space Almanac. New York: World Almanac.

Pelton, J.N. (1998) "Telecommunications for the 21st century". Scientific American. April 1998. Volume 278, Number 4, page 80-85.

Rzepa, H. (1997) "Evolution in the digital jungle" in Chemical & Engineering News. November 24, 1997 Volume 75 Number 47, pages 71-74.

Seifert, H.S. (1959) Technology. New York: John Wiley and Sons, Inc.

Shipman, H.L. (1987) Space 2000: Meeting the Challenge of a New Era. New York: Plenum Press.

Smith, D.D. (1978) Teleservices Via Satellite: Experiments and Future Perspectives. Boston: Sijthoff & Noordhoff.

Chapter 2

PROCEEDINGS OF SPIE

SPIE – The International Society for Optical
Engineering

*Free-Space Laser
Communication Technologies XII*

G. Stephen Mecherle
Chair/Editor

24 January 2000
San Jose, California

Sponsored and Published by
SPIE – The International Society for Optical
Engineering

Volume 3932

SPIE is an international technical society
dedicated to advancing engineering and scientific
applications of optical, photonic, imaging,
electronic, and optoelectronic technologies.

SESSION 3
System Design and Analysis

Tripartite Systems: Wireless Radio Frequency, Free
Lasers and Wired Terrestrial Systems
Integration

Bradley S. Tice

Advanced Human Design, Cupertino, CA 95015-
2214

ABSTRACT

Because of the use of radio frequency, free
space lasers and wired terrestrial communication
systems are often separate, and at times conflicting
systems, a proposed unifying system of all three
modes of receiving and transmitting will be
addressed under the collective title of Tripartite

Systems. Because of the vast network of wired terrestrial communication systems in the Western world communications systems of the future will still use wired systems as the backbone of signal transmission for most data. Such a network does have limits, i.e. the need to replace existing copper wire with fiber optics, but with the advent of free space lasers, lasecom systems, that have bit-error rates at acceptable levels for commercial and military applications, along with radio frequency systems, will give wired systems a 'complimentary' competition that will improve the overall efficiency of modern communication systems.

1. INTRODUCTION

The key to using radio frequency and free space lasers and wired terrestrial systems is how it will be integrated. Being that radio frequency and free space lasers allow for 'wireless' transmission and reception they can be used for space and atmospheric level communication nodes. Satellites for space and HALE's (High Altitude Land Endurance) vehicles that can be used as airborne networks, or AirNets, between space and terrestrial communication relays (See Figures 1, 2 and 3). Increasing the speed of wired systems will come with the mass use of fiber optics to replace the limited speed of existing copper wire networks. Integrated communication systems will make wireless and wired communication both seamless and efficient for the 21st Century.

1.1 Theory

In 1993 Nicholas Negroponte of MIT suggested that the future of telecommunications would be a large 'flip-flop' in that narrowband services that are now being carried by glass fiber would transfer to wireless operations and broadband services from radio waves and satellites to that of fiber optics and coaxial cable (Pelton, 1998:82). Pelton, in what has been labeled the 'Pelton Merge' advocates and opposing view in that the future will feature a 'rich but confused' digital mixture of fiber, coaxial, terrestrial wireless and satellite services (Pelton, 1998:82). While the 'Pelton Merge' seems to be the more likely of the two theories to develop, such an integration of various wired and wireless systems need not be 'confused' nor wholly digital, but a seamless interconnection between such systems is a prerequisite for a functioning integrated whole.

1.2 Satellites

In reading some general secondary works on satellites it becomes clear that even as late as twenty years ago the idea of using satellites to relay all types of data around the world was being considered. Martin points out in his The Wired Society (1978) that the total investment of terrestrial telecommunications facilities is 'huge'. He states "AT&T alone is spending $9 to $10 billion per year on capital improvement of the Bell system" (Martin, 1978:149). He continues "The annual revenue from telecommunications in the United States is over $40 billion and is growing by about $5 billion per year. Much of the capital expenditure in the telecommunications industry is going into trunks and trunk switching that satellites and demand-assignment equipment could replace" (Martin, 1978:149).

Martin makes it clear that such uses can have tremendous cost savings. "If earth stations were associated with the toll offices in the 500 most populous cities, such a satellite network would cost less that $2 billion. If it had a lifetime of ten years, the cost would be $200 million per year. (Most telecommunications equipment has a forty-year lifetime.)" (Martin, 1978:150).

It is clear from the general investment in existing ground faculties, usually public works or private utilities companies that are paid for and maintained by tax dollars or direct charges, of the developing internet, such long term considerations were either ignored or forgotten when designing the current land based systems.

1.3 Telecommunications

With the rapid rise of a global economy, the explosive growth of the telecommunications industry will have a global impact in all aspects of our lives. In looking back a decade ago, the expected growth of international telephone traffic would double from the period 1983 to 1988 and double again from 1988 to 1993 (Shipman, 1987: 83).

With this increased demands comes an increase in efficiency of research, design, manufacture, and operation of such satellite systems. Because all communication satellites must be in the same geosynchronous orbit that circle the earth's equator once every twenty-four hours, there is only a limited number of satellites that can be placed in that orbit, and are regulated by the

World Administrative Radio Conference, WARC (Shipman, 1987:83).

The way technology is transferred from government research to private industry is through what is called Space Technology Integration or STI. As Smith explains from his <u>Via Satellite</u> (1978) "Space Technology Integration is a specific application of the basic TI framework which is relevant to the communication satellite experimental process. There are two major cycles in the process – the innovational and institutional cycles. The two cycles meet at the point where a specific application has reached the stage when it is ready for operational implementation. At this point the application goes on to the institutional cycle, and the innovational cycle begins again", (Smith, 1978:2).

Such a cycle can be seen with such organizations as NASA. In 1971 during a reorganization of its Office of Space Science and Applications, a new office was established, the Office of Applications (Smith, 1978:18). The central idea behind this new office was explained by then NASA's Administrator, James C. Fletcher. "The application of space technology to solving problems here on Earth is perhaps NASA's most important new thrust. I believe it appropriate to centralize into a single office in NASA Headquarters all of the resources which we can muster to support space applications to Earth" (Smith, 1978:18).

CURRENT DEVELOPMENTS

An overview of current technology gives a glimpse of the potential development of a satellite based distributional system for a global computer network. From the development of wireless modems that weigh but ounces to the rapid growth of satellites based mobile phones, the trend is towards a wireless world (Bluck, 1998:4).

One of the most important factors in driving this new market is the desire for more 'speed' in relation to a systems carrying capacity. Such a carrying capacity is a function of the power of the signal, the bandwidth of the signal, and the amount of noise that is present (Hogan, 1997:37). This is termed 'Shannon's Theorm' and determines how much data a medium can carry.

Megabit per second transmissions are only currently possible in three formats: cable, xDSL, and satellite (wireless). A trouble area for the development of a satellite Internet system is the problem of distance in regards to the Internet TCP/IP protocol that holds each sent packet until an acknowledgement is received of a successful transmission (Hogan, 1997:37). For a satellite in a geosynchronous orbit, 22,000 miles overhead, the delay for the round trip is half a second.

One solution is to drop the requirement for the 'acknowledgement' thus removing a extraneous part of the protocol. Another is by adding more memory to the buffer (Hogan, 1997:51). Deregulation of the radio frequencies have also contributed to developments such as local multipoint distribution service or LMDS that have a lower maintenance costs than copper or cable-based networks (Garber, 1998:14).

This cost savings also applies to satellite applications to the internet. There are 'substantial' savings in utilizing the Internet satellite technology as opposed to relatively expensive high-bandwidth terrestrial systems (Kilaski, 1997:52). This will be important in the near future as the internet is expected to grow from 50 million subscribers today to 150 million by the year 2000 with a growth of $30 billion in the ensuing years (Evans, 1998: 76). With the expected investment of around $400 billion for future Q/V-band systems, wave lengths of six to eight millimeters, that are technically more difficult than the currently proposed Ka-band that encompasses wavelengths between 1 and 1.5 centimeters that have serious attenuational problems with precipitation (Evans, 1998:76).

2.1 Error Rates for Satellite Paths

High error rates are of serious concern because 1.) they cause errors in datagrams which will necessitate retransmission and 2.) TCP usually interprets loss as a sign of congestion and returns to a modified version of a slow start (Partridge and Shepard, 1997:48). While there have been suggestions for quantifying acceptable error rates, i.e. view the TCP's natural frequency of congestion avoidance starts and then seek an error rate that is substantially less than that frequency, there is no really adequate answer to this dilemma (Partridge and Shepard, 1997:48).

2.2 Intelligent TCP

It has been put forward that in conjunction with reducing satellite error rates the development of 'intelligent' TCP that would handle transmission errors should be developed. These two approaches are 1.) Either the TCP can Explicitly be told that link errors are occurring or 2.) TCP can infer that link errors are occurring (Partridge and Shepard, 1997: 49). A third way is to deal with data information as an autonomous neutral transport system, i.e. the data itself is it's own governor and becomes the agent of qualification through a process of 'regulated' behavior, not unlike a living neural network system. All three systems will take some time to develop[1].

2.3 AirNets

Airnet is the exclusive use of HALE (High Altitude Long Endurance) platforms as a local network of airborne telecommunication nodes that function either as the sole non-terrestrial relay junction, a closed AirNet, or in conjunction with space, i.e. satellites, or 'temporary' airborne relays such as commercial or military aircraft or scientific balloons, that would make it an open AirNet system (See Figures 4 and 5). AirNets have the power to afford wireless telecommunications without expensive space satellites hence the term for HALEs as 'poorman's' satellites. AirNets can also be used as physical markers for 'control' of local airspace by implementing data encryption,

[1] I have developed Autonomous Neural Transport systems: ANTS in an unpublished theory paper.

specifically coded data signals, for both uplinks and downlinks within an AirNet network.

When discussing satellite orbits, there are three categories of orbits based on the distance from the earth. Closes to proximity to the earth is the LEO satellites, for Low Earth Orbit, that circle hundreds of kilometers above the surface of the earth. Next are the MEO satellites that represent a Medium Earth Orbit and at 36,000 kilometers out from earth are the GEO satellites (Pelton, 1998:81).

2.4 NETSAT

NETSAT's are networking satellites used to relay computer related data signals and form a neural network of wireless data transmissions that form a global network to local computer systems (Tice, 1998).

2.5 Dual Integrated Systems

The use of both radio frequency and optical systems, i.e. lasecom systems, in HALE's as dual integrated systems would provide optimal speed, optical, with optimal robustness of signal transmission, radio frequency, that could be alternated according to environmental dictates, i.e. weather conditions, while keeping costs at a minimum (see paper LA3932-14 this conference)

SUMMARY

Wireless radio frequency, free space lasers and wired terrestrial systems will integrate into a logical, cost effective manner that will place new technology along side existing technology to make for a collaborative effort, rather than a competitive one, that will drive communicative services into a seamless whole.

ACKNOWLEDGEMENT

I wish to thank NASA Dryden Flight Research Center in Edwards, California USA for the use of photographs on various HALE vehicles used in this paper.

REFERENCES

1.   J. Bluck, J., "Ames offers wireless modems slightly bigger than pagers" The Ames Astrogram,, March 20, pp. 4 & 8, 1998.

2. J.V. Evans, "New satellites for personal communications", Scientific American, 278, pp. 70-77, 1998.
3. L. Garber, "Auction sets stage for new broadband wireless service", Computer, 3 1, 1998.
4. H. Hogan, "Paving the information driveway", High Technology Careers Magazine, 14, pp. 19, 37, 51, 1997.
5. D. Kilarski, "Satellite networks: Data takes to the skies", Network, 12, pp. 52-58, 1997.
6. J. Martin, The Wired Society, Prentice Hall, Inc., Englewood Cliffs, 1978.
7. C. Partridge, and T. Shepard, "TCP/IP performance over satellite links", IEEE Network, September/October 1997, pp. 44-49, 1997.
8. J.N. Pelton, "Telecommunications for the 21st century", Scientific American, 278, pp. 80-85, 1998.
9. H.L. Shipman, Space 2000: Meeting the Challenge of a New Era. Plenum Press, New York, 1987.
10. D.D. Smith, Teleservices Via Satellite: Experiments and Future Perspectives, Sijthoff & Noordhoff, Boston, 1978.
11. B. Tice, "NETSAT: the use of telecommunication satellites for networking

computer systems". Paper presented at the 1998 AIAA/AAS Astrodynamics Specialist Conference. Boston Park Plaza, Boston, Massachusetts. August 10-12, 1998.

Bradley S. Tice

FIGURE 1

[HELIOS Prototype Takeoff]

FIGURE 2

[HELIOS Prototype in Flight]

FIGURE 3

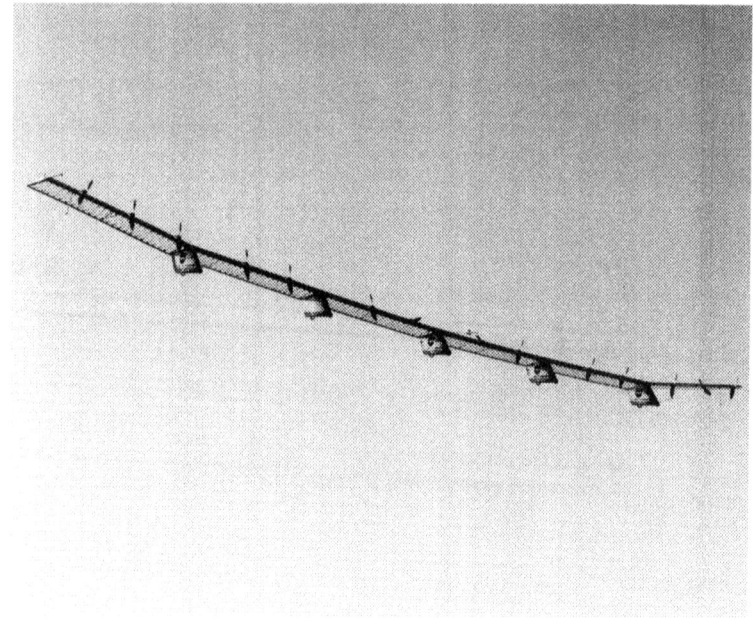

[HELIOS Flying Wing]

FIGURE 4

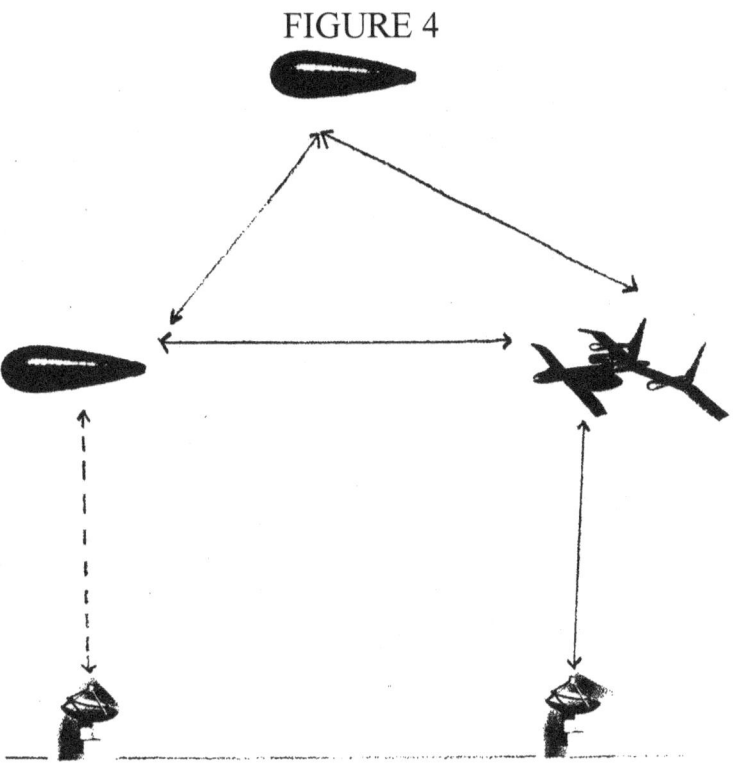

[Closed AirNet System]

FIGURE 5

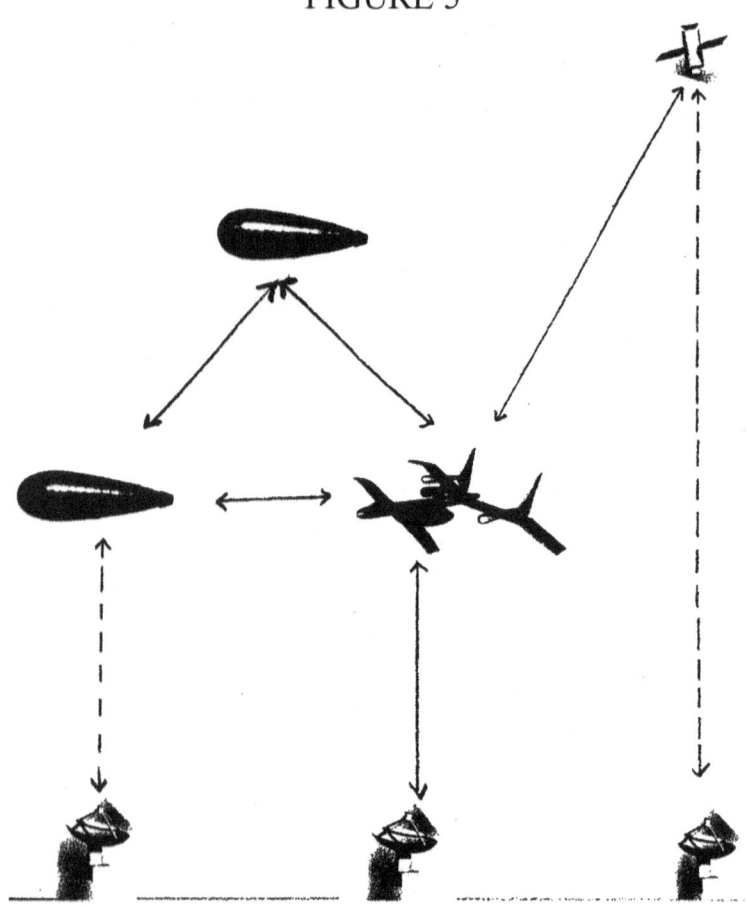

[Open AirNet System]

Chapter 3

PROCEEDINGS OF SPIE

SPIE – The International Society for Optical Engineering

Free-Space Laser Communication Technologies XII

G. Stephen Mecherle
Chair/Editor

24 January 2000
San Jose, California

Sponsored and Published by
SPIE – The International Society for Optical Engineering

Volume 3932

SPIE is an international technical society dedicated to advancing engineering and scientific applications of optical, photonic, imaging, electronic, and optoelectronic technologies.

SESSION 3
System Design and Analysis

Dual Integrated Systems for HALE

Bradley S. Tice

Advanced Human Design, Cupertino, CA 95015-2214

ABSTRACT

Because free lasers offer faster transmittal of data than radio frequency, but are less reliable because of cloud cover or fog, a two stage system can be developed using HALE, high altitude long endurance, platforms as a transitional link between space and terrestrial communication nodes. Because HALE platforms will fly above cloud layers, 35,000 feet, such platforms could use either radio frequency or free space lasers for 'air' to 'ground' transmissions and free lasers for 'space' to 'air' phase of transmissions. The use of radio frequency or free laser would depend on the type of environment during operations. Vehicles that are remotely operated, solar powered platforms that function at 100,000 feet indefinitely.

With laser communication field research supporting the successful transmission of a high signal quality, bit-error rate or BER, in a line of sight operation that was obscured with heavy degradation elements such as heavy fog and smoke, that is applicable to both commercial and military applications. Such research has lead to the use of a bit-error rate of high bandwidth (570 mbits/ s) lasecom system that was correlated with the atmospheric transmission of 0.25% and a BER of less that 10 to the tenth power at an atmospheric transmission above 2.5%. System performance was about 10 db less than calibrated. There was no use of error correction schemes or active tracking to achieve these results.

When this technology is adapted for use in a high altitude long endurance vehicle (HALE) and complimented with existing radio frequency systems, a dual integrated system results that increases the reliability of signal transmissions from terrestrial to space, and transmission relays back, with HALE being the 'environmental' filter that would utilize either free space laser or radio frequency depending on the line of sight conditions. A HALE platform, named HELIOS, has recently been put on the market that has a wing span of 247 feet and can stay aloft at 100,000 feet indefinitely. Designed by NASA, HELIOS is being build by private industry.

1. INTRODUCTION

Dual Integrated Systems is a systems oriented application of a two stage, i.e. two separate systems that function alternately, employing two different wireless transmission systems to terrestrial communication nodes and a single, i.e. either radio frequency or optical system, to space communication nodes. The platform for this Dual Integrated System, or DIS, is HALE; High Altitude Long Endurance, vehicles that are remotely operated, solar powered platforms that function at 100,000 feet indefinitely.

1.1   Dual Integrated Systems

The Reason for a dual integrated system for an airborne networking node, or AirNet, is really a function of systems reliability and the general robustness of signal transmission and reception for both commercial and military applications. While radio frequency, or RF, is a reliable and robust method of wireless transmission, it suffers from being a slower method of signal communication than optical systems, i.e. lasers, and as such is a limiting criteria for the use of radio frequency in future high speed data signal environments. While optical systems, lasers, can be used for communication systems, i.e. lasecom, they suffer when the signal is exposed to heavy degradation elements such as heavy fog and smoke. While current field tests have shown that it is possible to achieve a bit-error rate, BER, that is acceptable for a reliable and robust system of communication that would be acceptable for both commercial and military applications, the effects of scintillation are still a limiting factor for all environmental conditions.

1.2   HALE

HALE's are High Altitude Long Endurance vehicles that are vehicles that are remotely operated, solar powered platforms that function at 100,000 feet indefinitely. NASA has developed them, in a chronological order, Pathfinder Plus, Centurion, and HELIOS (See Figures 1-5). HELIOS is the commercial version of Centurion and is currently on the market for around $3 to 5 millions dollars per craft (San Jose Mercury Newspaper, 1999: 7B) (See Figure 7). These vehicles will provide the airborne transport of the Dual Integrated Systems as at 100,000 feet they are well above the commercial flights and environmental hazards, i.e. smoke, fog and heavy clouding, that interrupt line of sight optical operations. HALE's can be made the mainstay of Dual Integrated Systems and like their lighter than air counter parts, i.e. dirigibles, they are remotely controlled and fly indefinitely at 100,000 feet AGL.

3.3   Phase-Array Antennas

Current studies have indicated that HALE platforms, High Altitude Long Endurance, could support phase-array antennas with up to 3500 beams that could achieve video signal transmission distribution points up to 500 kilometers apart and facilitate two way mobile communications (Pelton, 1998:85). While this is a successful implementation of radio frequency technology, the real question is not the amount of beams or signals but the speed of that signal in a beam or beams. Phase-array antennas are a temporary fix to wireless airborne situations but they do not address the growing need for speed that only optical wireless options, i.e. lasecom or laser communication, systems promise. This is why Dual Integrated Systems provide the speed and robustness found in optical and radio frequency dual applications.

4.4   Bit-Error Rates

With laser communication field research supporting the successful transmission of a high signal quality, bit-error rate or BER, in a line of sight operation that was obscured with heavy degradation elements such as heavy fog and smoke, that is applicable to both commercial and military applications. Such research has lead to the use of a bit-error rate of high bandwidth (570mbits/s) lasecom system that was correlated with the atmospheric transmission over a folded path of 2.4 km. This resulted in BER's of 10 to the 7th power at an atmospheric transmission of 0.25% and a BER of less that 10 to the tenth power at an atmospheric transmission above 2.5%. System performance was about 10 db less than calibrated. There was no use of error correction schemes or active tracking to achieve these results (Strickland, Lavan, and Woodbridge, 1999:424). When this technology is adapted for use in a high altitude long endurance vehicle (HALE) and complimented with existing radio frequency systems, a dual integrated system results that increases the reliability of signal transmissions from terrestrial to space, and transmission relays back, with HALE being the 'environmental' filter that would utilize either free space laser or radio frequency depending on the line of sight conditions.

SUMMARY

The development and commercial sales of HALE vehicles signals an ideal platform for integrating systems such as Dual Integrated Systems (DIS) into select markets for testing purposes as with its joint robustness (Radio Frequency) and optimal signal transmission speed (lasecom systems) an efficient and cost effect system of local networks (AirNets) could revolutionize commercial and military telecommunication applications. With the advent of the internet the need for fast data transmission makes this system even more time sensitive than when I first presented my ideas on satellite telecommunications of the future (Tice, 1998). Dual Integrated Systems is the airborne wave of future telecommunications and represents optimal integrated planning for both the wired and wireless world.

ACKNOWLEDGEMENT

I wish to thank the NASA Dryden Flight Research Center at Edwards, California USA for the use of materials and photographs of various HALE vehicles used for this paper.

REFERENCES

1. J.N. Pelton, "Telecommunications for the 21st century", Scientific American, 278, pp. 80-85, 1998.
2. San Jose Mercury Newspaper, "On auto pilot", Thursday, October 14, 1999, pp. 7B.
3. M. Seldon, "Lasers pinpoint satellite locations", Aerospace America, October 1999, pp. 36-39.
4. B.R. Strickland, M.J. Lavan, and E. Woodbridge, "Effects of fog on the bit-error rate of a free-space laser communication system", Applied Optics, 38, pp. 424-431.
5. B. Tice, "NETSAT: the use of telecommunication satellites for networking computer systems", Paper presented at the 1998 AIAA/AAS Astrodynamics Specialist Conference. Boston Park Plaza, Boston, Massachusetts, August 10-12, 1998.

FIGURE 1

[HELIO Prototype Takeoff]

FIGURE 2

[Centurion]

FIGURE 3

[Centurion Flying Wing]

FIGURE 4

[Pathfinder Plus]

FIGURE 5

[Pathfinder]

FIGURE 6

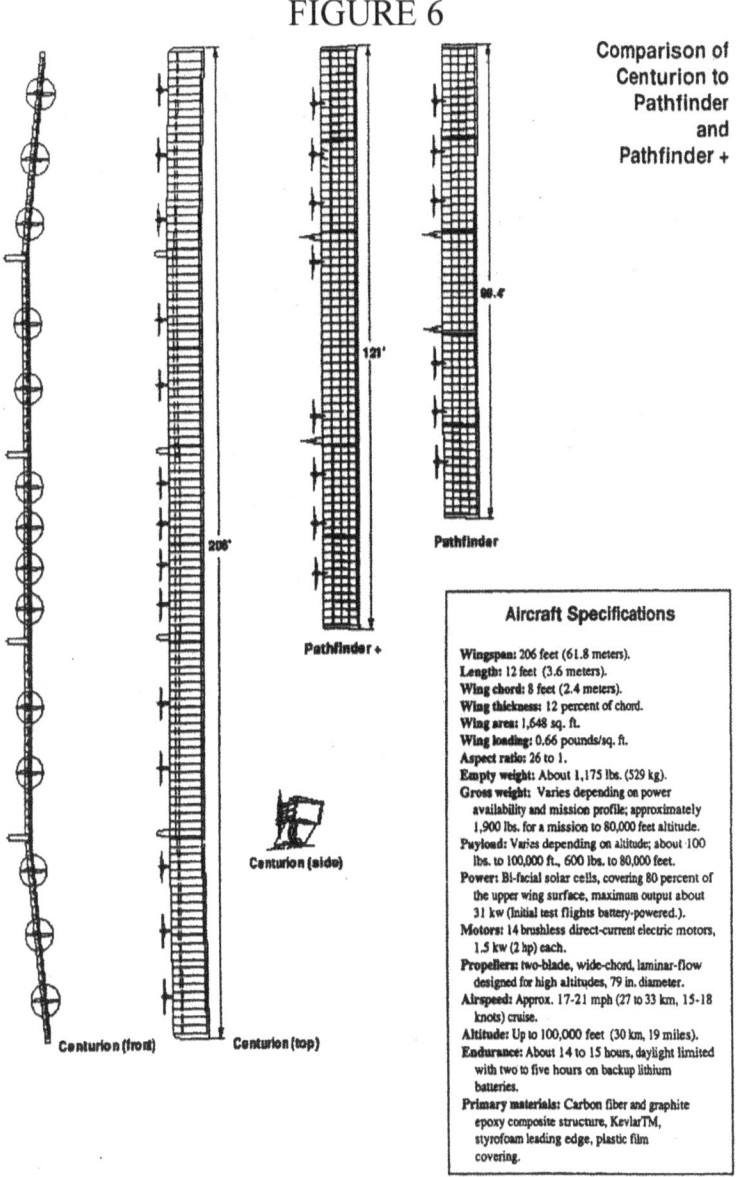

Comparison of
Centurion to
Pathfinder
and
Pathfinder +

Aircraft Specifications

Wingspan: 206 feet (61.8 meters).
Length: 12 feet (3.6 meters).
Wing chord: 8 feet (2.4 meters).
Wing thickness: 12 percent of chord.
Wing area: 1,648 sq. ft.
Wing loading: 0.66 pounds/sq. ft.
Aspect ratio: 26 to 1.
Empty weight: About 1,175 lbs. (529 kg).
Gross weight: Varies depending on power
  availability and mission profile; approximately
  1,900 lbs. for a mission to 80,000 feet altitude.
Payload: Varies depending on altitude; about 100
  lbs. to 100,000 ft., 600 lbs. to 80,000 feet.
Power: Bi-facial solar cells, covering 80 percent of
  the upper wing surface, maximum output about
  31 kw (initial test flights battery-powered.).
Motors: 14 brushless direct-current electric motors,
  1.5 kw (2 hp) each.
Propellers: two-blade, wide-chord, laminar-flow
  designed for high altitudes, 79 in. diameter.
Airspeed: Approx. 17-21 mph (27 to 33 km, 15-18
  knots) cruise.
Altitude: Up to 100,000 feet (30 km, 19 miles).
Endurance: About 14 to 15 hours, daylight limited
  with two to five hours on backup lithium
  batteries.
Primary materials: Carbon fiber and graphite
  epoxy composite structure, KevlarTM,
  styrofoam leading edge, plastic film
  covering.

[Centurion and Pathfinder Specifications]

FIGURE 7

On auto pilot

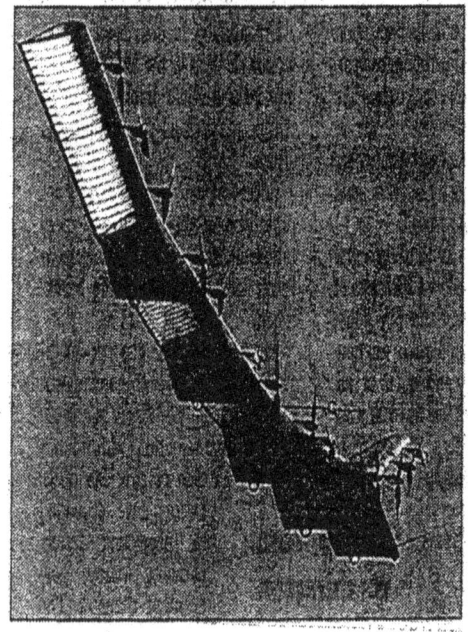

Helios, left, an unmanned aircraft developed by NASA and built by private industry for sale to the public, was put on the market Wednesday in an effort by the space agency to show a practical application to its research. The aircraft, which has a 247-foot wingspan, will sell for around $3 million to $5 million, depending on the the options. Called the "poor man's satellite" because of its relatively low cost, the Helios is propelled by 14 electric motors fueled by the sun and can stay aloft at 100,000 feet indefinitely. Researchers and manufacturers say the aircraft performs many of the same functions as communication and imaging satellites.

ASSOCIATED PRESS

[HELIOS]

Summary

Again, as outlined in the introductory chapter of this dissertation, each paper stands as a single supported concept, or groups of concepts, that also combines aspects of the other papers to the notions found in a systematic approach to general telecommunication theory. The result of these papers, besides as a showcase for my technical knowledge and engineering vision, is a practical application of telecommunication theory to local, national, and global aspects of telecommunications.

About The Author

Dr. Tice is the founder and CEO of Advanced Human Design located in Cupertino, California U.S.A. Dr. Tice has been an Ames Associate at the NASA Ames Research Center at Moffett Field, California U.S.A. and he is currently a Mission Specialist Astronaut Candidate with NASA.